LARISA   ALTSHULER

JOURNEY THROUGH GEOMETRY

LOGIC

PLANE GEOMETRY

SOLID GEOMETRY

COORDINATE   GEOMETRY

VECTORS

TRANSFORMATION GEOMETRY

PROJECTIVE GEOMETRY

NON-EUCLIDEAN GEOMETRIES

To Louiza, Jennifer, and Daniel – my wonderful talented grandchildren who took course of Geometry in middle school

CONTENTS

INTRODUCTION

Hello everyone who is holding this small book in your hands. This is not a textbook and it cannot replace one. As the name suggests, this book contains a brief reference of main facts you need to know and remember after a course of Geometry. The facts are logically ordered and systematically presented for the ease of understanding, memorization and subsequent lookup. We neither explain how the facts are obtained, nor discuss how they can be used. We do not prove any theorems. All of that you can find in your textbook. Our goal for this book is to help students sort things out *after* they took Geometry, or any unit of it. If you are presently a student of Geometry, or need to dust off the old knowledge, this book will give you the overview of the course, refresh most important definitions, theorems and formulas of Geometry, and enable you to solve geometrical problems. We believe that the brief will be very helpful for you to successfully prepare for final exam, proficiency exam or higher level courses.

One of the reason why you need to remember and understand the fundamental geometric facts is that Geometry is interrelated with many other parts of science, like Algebra, Calculus, and Physics, to name a few. Sometimes it is even hard to classify what belongs to where. Consider for instance **sin** or **cos** theorems. Are they part of Geometry or Trigonometry? The knowledge of triangles in Geometry prepares you to understand the trigonometric formulas. Similarly, the knowledge of the formulas for the area of circle or the volume of sphere covered in Geometry prepares you to succeed in Calculus. On the other hand, to solve the geometrical problem of finding a point common to three planes, one often needs to reserve to Algebra and solve a system of three linear equations.

In this book we mainly collected geometric facts which are extremely important and can be easily remembered. If the formulas are interesting but not widely used we mark them by an astcrisk.

For those of you who are well familiar with other books on Geometry, we need to say a few words about the definitions and notation used in this book. To reach the complete clearness and intuitive understanding of the geometric facts , we allowed ourselves some leeway in a few definitions. For instance, the definition of the polyhedron we gave in this book is not precise, but sufficient for quick and clear understanding. In this book we adopted non-traditional, but in our opinion very convenient notation designed to stress intuitive comprehension over formal correctness. For example, a segment, a straight line, a ray, and the measure of the segment are all denoted identically by the names of two points they pass through.

If we write **AB⊥a**, we mean that the segment (or straight line, or ray) **AB** is perpendicular to the line a. If we write **AB = a**, we mean that the length of the segment **AB** is equal to **a** units. Our teaching experience has convinced us that such notations are completely understandable by context and can never be a reason for misinterpretation of information about geometric figures. Let us give one more example of the same approach to notation: $\angle$**ACD** is formed by two chords. Angle **ACD** is the geometric figure. In the book we state that $\angle$**ACD** = ½ ($\cup$**AE**-$\cup$**BD**), which means "the measure of an angle **ACD** is equal to ½ of the difference of measures of arcs **AE** and **BD** ". Using the traditional notation, we would have to write **m**$\angle$**ACD** = ½ (**m**$\cup$**AE** −**m**$\cup$**BD**), which we find to be longer and unnecessary.

In conclusion we would like to mention that Geometry is one of the greatest parts of science. Learning Geometry we rediscover again the facts people knew many centuries (even millenniums) ago. At the same time, the science of Geometry has made an amazing progress throughout the history of mankind, from discovery of simple properties of figures on the plane to understanding of complicated forms in multidimensional spaces. After one course in Geometry, to which this book is devoted, we only open a small window into the world of GEOMETRY. As any teacher, we are delighted and gratified when some of our students say in the end: "Now I know something about it. What is ahead?"

LOGIC
TABLE 1

	DEFINITION	NOTATION
Statement	**Statement** is a declarative sentence that is true or false but not both.	p,q,...
Negation	**Negation** is a new statement that has opposite to original statement truth value.	~p,~q,..
Conjunction	**Conjunction** is a compound statement involving logical connective "**and**".	p∩q
Disjunction	**Disjunction** is a compound statement involving logical connective "**or**".	p∪q

TABLE 2
CONDITIONAL STATEMENTS

Implication	**Implication** is a compound statement involving logical connective "**if-then**".	$p \rightarrow q$
Converse	**Converse** is an original statement where the hypothesis and conclusion are replaced.	$q \rightarrow p$
Inverse	**Inverse** is an original statement where the hypothesis and conclusion are substituted by their negations.	$\sim p \rightarrow \sim q$
Contrapositive	**Contrapositive** is a converse statement where the hypothesis and conclusion are substituted by their negations.	$\sim q \rightarrow \sim p$

TABLE 3
LAWS OF LOGIC

Law of Detachment	Direct Reasoning	$p \rightarrow q$ p $\therefore \quad q$
Law of Syllogism	Chain Rule	$p \rightarrow q$ $q \rightarrow r$ $\therefore \quad p \rightarrow r$
Law of Contraposition	Indirect Reasoning	$p \rightarrow q$ $\sim q$ $\therefore \quad \sim p$

TABLE 4

AXIOMATIC SYSTEM

THEOREMS
Axioms and Postulates
Definitions

Undefined Terms

$\updownarrow$

Point, straight line, plane, set

TABLE 5
POSTULATES

1.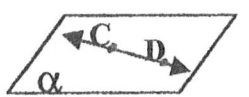

   Each line contains at least two points.

2.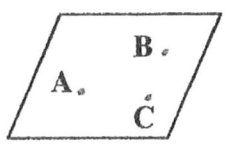

   Through any two points, there is exactly one line.

3.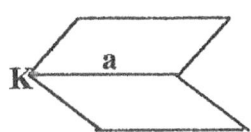

   If two points lie in a plane, then the line joining them lies in that plane.

4.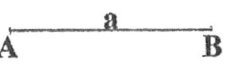

   A plane contains at least three noncollinear points.

5.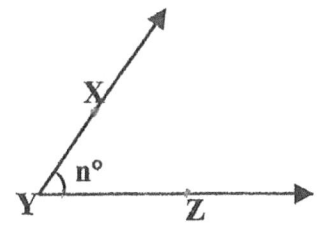

   If two planes have a common point, then they have at least a common line.

6.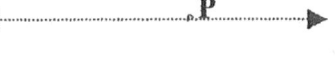

   Each segment corresponds to the unique real number described its length.

   $$AB = a > 0$$

7.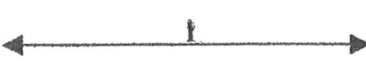

   Each angle corresponds to the real number described its measure.

   $$\angle XYZ = n° \geqslant 0$$

8.

   Parallel Postulate (Playfair's Axiom)

   Given a line l and a point P not on l, there is only one line n containing P that is parallel to l.

TABLE 6
ANGLES

$\angle ABC = 90° = \frac{\pi}{2}$ rad

Right Angle

$\alpha < 90°$

Acute Angle

$90° < \beta < 180°$

Obtuse Angle

$180°$

$\angle AOB = 180°$

Straight Angle

Complementary Angles

$\alpha + \beta = 90°$

Supplementary Angles

$\alpha + \beta = 180°$

VERTICAL ANGLES

THEOREM

Vertical angles are congruent

$\angle 1 = \angle 3 , \quad \angle 2 = \angle 4$

ANGLES FORMED BY TWO LINES AND TRANSVERSAL

$\angle 3, \angle 4, \angle 5, \angle 6$ are interior
$\angle 1, \angle 2, \angle 7, \angle 8$ are exterior

$\angle 4$ and $\angle 6, \angle 3$ and $\angle 5, \angle 1$ and $\angle 7,$
$\angle 2$ and $\angle 8$ are alternate

$\angle 4$ and $\angle 5, \angle 3$ and $\angle 6, \angle 1$ and $\angle 8,$
$\angle 2$ and $\angle 7$ are one-side angles

$\angle 1$ and $\angle 5, \angle 2$ and $\angle 6, \angle 4$ and $\angle 8,$
$\angle 3$ and $\angle 7$ are corresponding angles

Property of Bisector

Any point on the bisector of an angle is equidistant from the sides of the angle.

$MA = MC$

TABLE 7
PARALLEL LINES

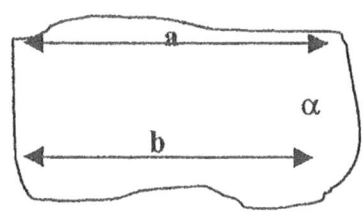

DEFINITION | Two lines are parallel if they belong the same plane and do not intersect.

$$a \parallel b \Leftrightarrow \{ a \cap b = \varnothing; \ a \subset \alpha, \ b \subset \alpha \}$$

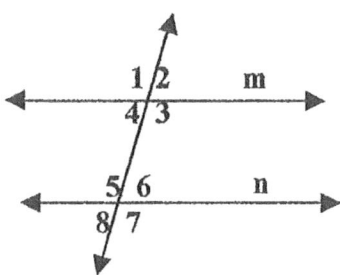

THEOREM | Two lines are parallel if and only if either alternate angles are congruent or corresponding angles are congruent, or the sum of one-side angles is 180°.

$$m \parallel n \Leftrightarrow \{ \angle 4 = \angle 6, \text{ or } \angle 2 = \angle 6, \text{ or } \angle 4 + \angle 5 = 180° \}$$

TRANSITIVE PROPERTY

If a∥b and a∥c , then b∥c.

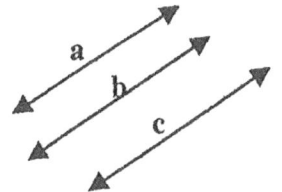

TABLE 8
PERPENDICULAR LINES

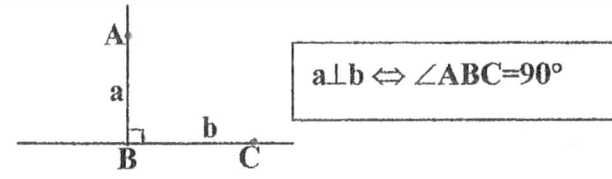

$$a \perp b \Leftrightarrow \angle ABC = 90°$$

Two lines are perpendicular if they form right angle.

PA⊥a, PA is a distance from point P to the line a.

PROPERTY OF PERPENDICULAR BISECTOR TO SEGMENT

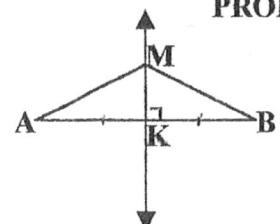

Any point on perpendicular bisector of a segment is equidistant from the ends of the segment.

$$\{ KA = KB, \ MK \perp AB \} \Leftrightarrow MA = MB$$

TRIANGLES

TABLE 9
ANGLES AND SIDES

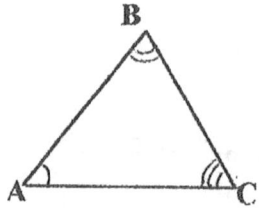

THEOREM

The sum of interior angles in a triangle is equal to 180°

$$\angle A + \angle B + \angle C = 180°$$

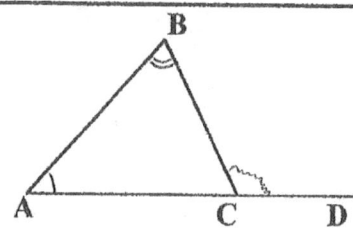

THEOREM

An exterior angle of a triangle is equal to the sum of the two nonadjacent interior angles.

$$\angle BCD = \angle A + \angle B$$

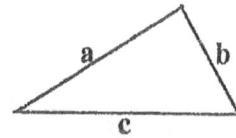

Triangle Inequality

$$|b-c| < a < b + c$$

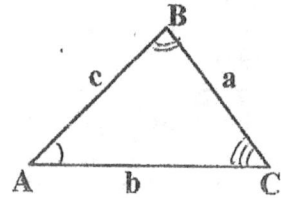

THEOREM

In a triangle opposite to a larger side there is a larger angle. Vise versa , opposite to a larger angle there is a larger side.

If $a > b > c$, then $\angle A > \angle B > \angle C$
If $\angle A > \angle B > \angle C$, then $a > b > c$

TABLE 10
CONGRUENT TRIANGLES

DEFINITION

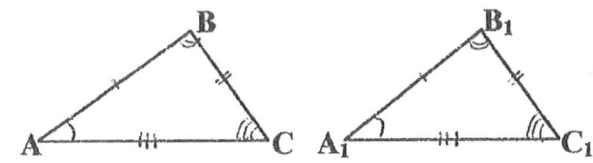

$$\Delta ABC \cong \Delta A_1 B_1 C_1 \Leftrightarrow \begin{array}{l} AB = A_1 B_1 \; ; \; \angle A = \angle A_1 \\ BC = B_1 C_1 \; ; \; \angle B = \angle B_1 \\ AC = A_1 C_1 \; ; \; \angle C = \angle C_1 \end{array}$$

SAS	ASA (AAS)	SSS
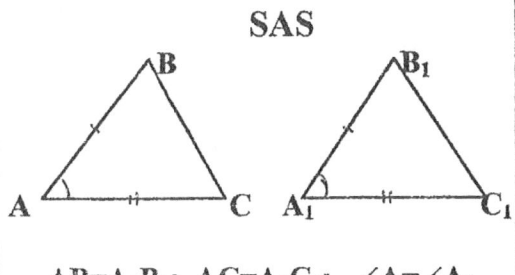	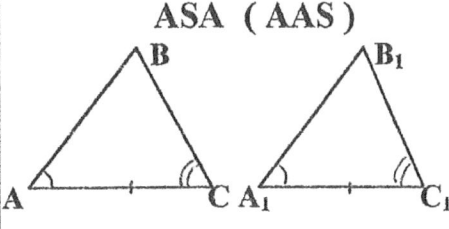	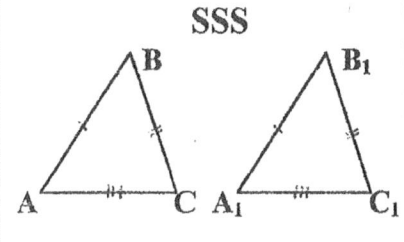
$AB = A_1 B_1$; $AC = A_1 C_1$; $\angle A = \angle A_1$	$\angle A = \angle A_1$; $\angle C = \angle C_1$; $AC = A_1 C_1$	$AB = A_1 B_1$; $BC = B_1 C_1$; $AC = A_1 C_1$

TABLE 11
MEDIAN OF A TRIANGLE

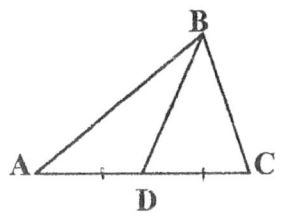

DEFINITION

The median in a triangle is a segment joining a vertex with a midpoint of its opposite side.

$AD = DC$, BD is a median.

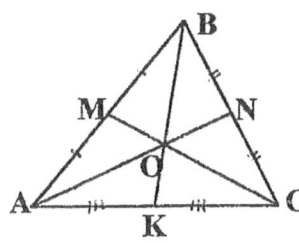

PROPERTIES OF MEDIANS

In any triangle the three medians intersect in the same point

O is a point of intersection of the medians (the center of gravity of the triangle).

$$\frac{AO}{ON} = \frac{BO}{OK} = \frac{CO}{OM} = \frac{2}{1}$$

11

TABLE 12
BISECTOR OF A TRIANGLE

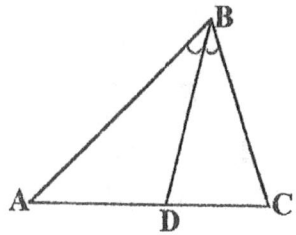

DEFINITION

The bisector of a triangle is a segment that divides the interior angle by two equal parts.

BD is a bisector, ∠ABD=∠CBD

PROPERTIES OF BISECTORS

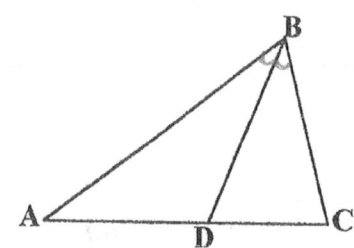

All three bisectors intersect in the same point.

O is a point of intersection of the bisectors (the center of inscribed circle or "incenter")

The bisector divides the opposite side by the segments proportional to the adjacent sides of the triangle.

$$\frac{AD}{DC} = \frac{AB}{BC}$$

TABLE 13
ALTITUDE OF A TRIANGLE

DEFINITION

BD is an altitude; BD⊥AC

An altitude of a triangle is the perpendicular segment from the vertex to the straight line containing the opposite side of the triangle.

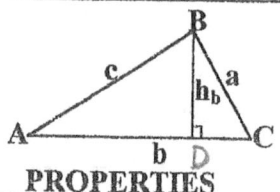

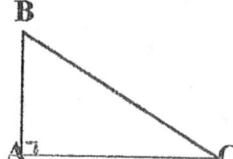

BA is an altitude; BA⊥AC

PROPERTIES

$$h_a : h_b : h_c = \frac{1}{a} : \frac{1}{b} : \frac{1}{c}$$

Straight lines containing the altitudes intersect in one point(orthocenter)

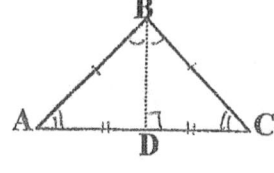

AB = BC

Isosceles Triangles

In an isosceles triangle the altitude to the base is a median and a bisector.

TABLE 14
RELATIONSHIPS BETWEEN THE ELEMENTS IN RIGHT TRIANGLE

$\angle C = 90°$; AC and CB are legs; AB is a hypotenuse.

CB = a; AC = b; AB = c; $\angle A = \alpha$; $\angle B = \beta$

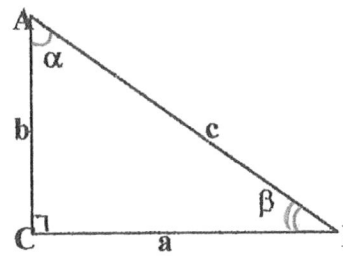

PYTHAGOREAN THEOREM

$$a^2 + b^2 = c^2$$

DEFINITION

$$\sin\alpha = \frac{a}{c}; \quad \sin\beta = \frac{b}{c}$$

$$\cos\alpha = \frac{b}{c}; \quad \cos\beta = \frac{a}{c}$$

$$\tan\alpha = \frac{a}{b}; \quad \tan\beta = \frac{b}{a}$$

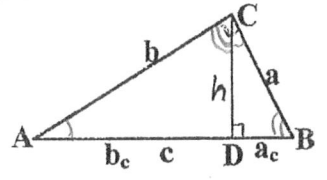

CD is an altitude

$\triangle ACD \sim \triangle ABC$

$\triangle CBD \sim \triangle ABC$

$\triangle ACD \sim \triangle CBD$

$$h^2 = a_c \cdot b_c$$
$$a^2 = c \cdot a_c$$
$$b^2 = c \cdot b_c$$

Particular Cases of Right Triangles

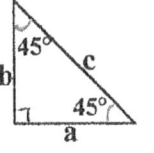

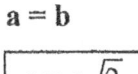

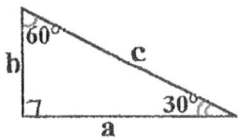

a = b

$$c = a\sqrt{2}$$

b = ½ c

$$A = b\sqrt{3}$$

TABLE 15
RELATIONSHIPS BETWEEN SIDES AND ANGLES IN A TRIANGLE

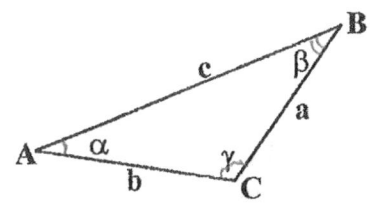

Sine THEOREM

$$\frac{a}{\sin\alpha} = \frac{b}{\sin\beta} = \frac{c}{\sin\gamma} = 2R$$

where R is a radius of circumscribed circle.

Cosine THEOREM

$$c^2 = a^2 + b^2 - 2ab\cos\gamma$$

COROLLARIES

1. If $c^2 = a^2 + b^2$, then $\gamma = 90°$, i.e. the triangle is right (Converse Pythagorean Theorem).

2. If $c^2 < a^2 + b^2$, then $\gamma < 90°$, i.e. the triangle is acute.

3. If $c^2 > a^2 + b^2$, then $\gamma > 90°$, i.e. the triangle is obtuse.

TABLE 16
SIMILAR TRIANGLES

DEFINITION

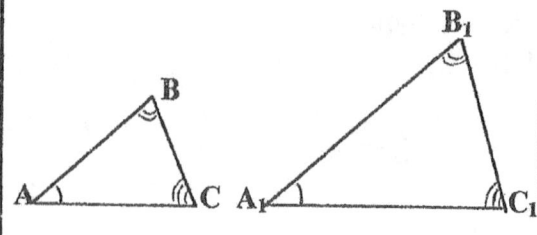

$$\Delta ABC \sim \Delta A_1B_1C_1 \quad \Leftrightarrow$$

$$\angle A = \angle A_1 ; \angle B = \angle B_1 ; \angle C = \angle C_1$$

$$\frac{AB}{A_1B_1} = \frac{BC}{B_1C_1} = \frac{AC}{A_1C_1}$$

TESTS FOR SIMILARITY

AA (AAA)	SAS	SSS

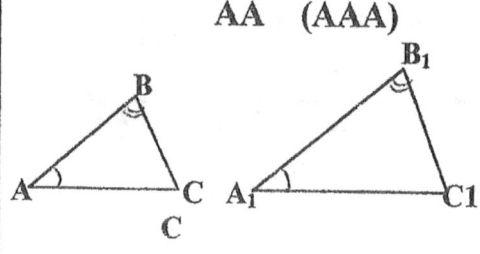

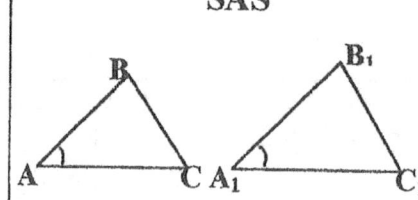

$$\angle A = \angle A_1 ; \quad \angle B = \angle B_1$$

$$\frac{AB}{A_1B_1} = \frac{AC}{A_1C_1} ; \quad \angle A = \angle A_1$$

$$\frac{AB}{A_1B_1} = \frac{BC}{B_1C_1} = \frac{AC}{A_1C_1}$$

COROLLARY

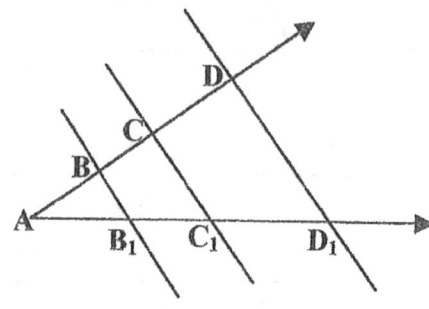

Parallel lines cut the sides of an angle by proportional pieces

$$\frac{AB}{AB_1} = \frac{BC}{B_1C_1} = \frac{CD}{C_1D_1}$$

Particular Case (THALES' THEOREM)

If $BC = CD$, then $B_1C_1 = C_1D_1$

MIDSEGMENT THEOREM

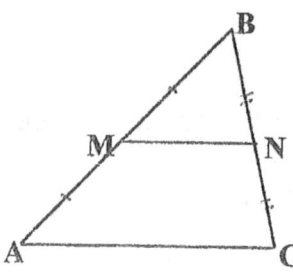

Midsegment of a triangle is parallel to the base and is equal to ½ of it.

$$MN \parallel AC ; \quad MN = \tfrac{1}{2} AC$$

QUADRILATERALS
TABLE 17
PARALLELOGRAM AND ITS PARTICULAR CASES

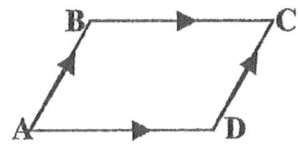

DEFINITION

A parallelogram is a quadrilateral with two pairs of parallel sides.

ABCD is a parallelogram	$\Leftrightarrow$	AB$\parallel$DC ; BC$\parallel$AD

PROPERTIES

1. Opposite sides of a parallelogram are congruent.

ABCD is a parallelogram if and only if AB=CD and BC=AD.

2. Opposite angles of a parallelogram are congruent.

ABCD is a parallelogram if and only if $\angle A=\angle C$ and $\angle B=\angle D$.

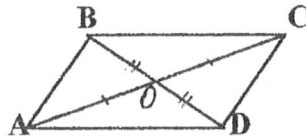

3. The diagonals of a parallelogram bisect each other.

ABCD is a parallelogram if and only if AO=OC and BO=OD.

4. The triangles formed by the diagonals have the same area.

Area($\triangle$AOB)=Area($\triangle$BOC)=Area($\triangle$COD)=Area($\triangle$AOD).

5. The sum of squares of the diagonals is equal to the sum of squares of the sides.

$$AC^2 + BD^2 = AB^2 + BC^2 + CD^2 + AD^2$$

RECTANGLE
DEFINITION

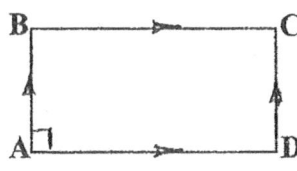

A rectangle is a parallelogram with all right angles.

PROPERTIES

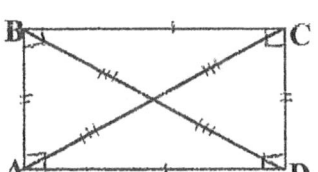

1. A rectangle has all properties of a parallelogram.

2. The diagonals of a rectangle are congruent.

If ABCD is a rectangle, then AC = BD

RHOMBUS

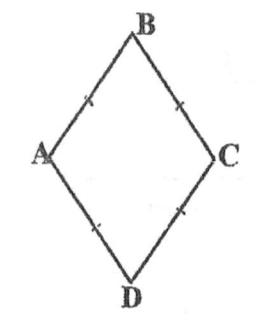

DEFINITION

> A rhombus is a parallelogram with all congruent sides.

PROPERTIES

1. A rhombus has all properties of a parallelogram.

2. The diagonals of a rhombus are perpendicular.

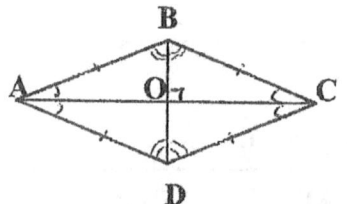

> If ABCD is a rhombus, then AC ⊥ BD

3. The diagonals bisect the opposite angles.

> If ABCD is a rhombus, then ∠BAO=∠DAO and ∠ABO=∠CBO

SQUARE

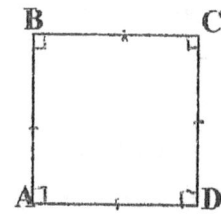

DEFINITION

> A square is a rectangle with all congruent sides.

OR

> A square is a rhombus with all right angles.

PROPERTIES

A square has all properties of a rectangle and a rhombus.

 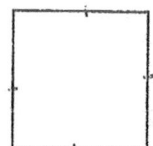 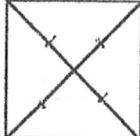 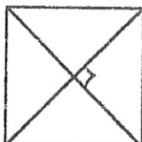 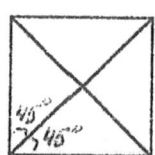

MIDQUAD THEOREM

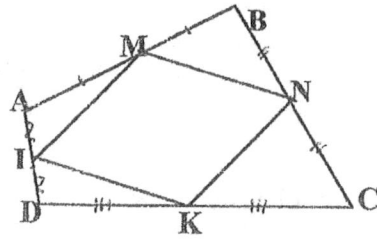

> If ABCD is a quadrilateral and M, N, K, L are
> midpoints of the sides, then MNKL is a parallelogram.

TABLE 18
TRAPEZOID

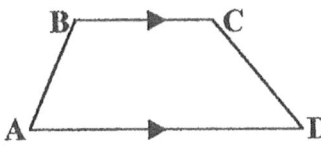

DEFINITION

A trapezoid is a quadrilateral with <u>only one pair</u> of parallel sides.

PROPERTIES

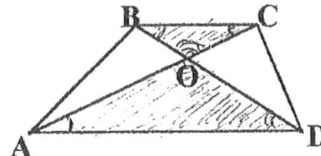

$$\Delta BOC \sim \Delta DOA$$

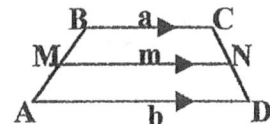

$$m = \frac{a+b}{2}$$

The diagonals of a trapezoid form two similar triangles.

Midsegment of a trapezoid is parallel to the bases and is equal to a half of its sum.

TABLE 19
KITE

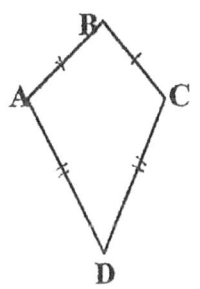

DEFINITION

A kite is a quadrilateral with two pairs of consecutive congruent sides.

AB=BC, CD=AD; ABCD is a kite.

PROPERTIES

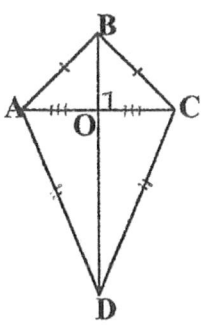

1. The diagonals of a kite are perpendicular.

$$AC \perp BD$$

2. One of the diagonal is bisected by another.

$$AO = OC$$

TABLE 20
POLYGONS

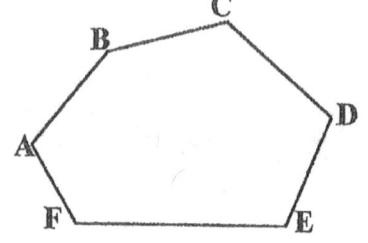

DEFINITION

A polygon is a closed simple plane figure bounded by segments.

ABCDEF is a hexagon

CONVEX POLYGON

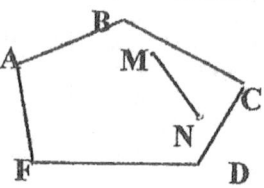

For any M and N
$M \in$ Polygon ,
$N \in$ Polygon, and
$MN \subset$ Polygon.

CONCAVE POLYGON

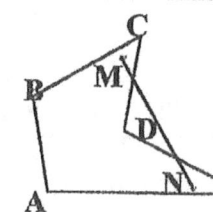

There exist such $M \in$ Polygon and $N \in$ Polygon , that $MN \not\subset$ Polygon.

THE SUM OF ANGLES OF n-gon

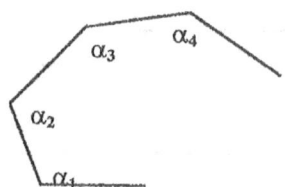

The sum of interior angles of a polygon is equal to $180°(n-2)$.

$$\angle\alpha_1 + \angle\alpha_2 + \angle\alpha_3 + \angle\alpha_4 + \ldots + \angle\alpha_n = 180°(n-2)$$

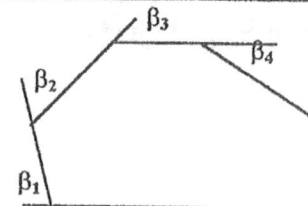

The sum of exterior angles of a polygon is equal to $360°$.

$$\angle\beta_1 + \angle\beta_2 + \angle\beta_3 + \angle\beta_4 + \ldots + \angle\beta_n = 360°$$

THE NUMBER OF THE DIAGONALS OF n-gon

A diagonal is a segment joining two not consecutive vertices.

The number of the diagonals $= \dfrac{n(n-3)}{2}$

REGULAR POLYGONS

Triangle

Square

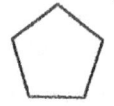
Pentagon

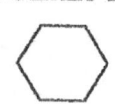
Hexagon

Octagon

A regular polygon is a polygon with congruent sides and congruent angles.

TABLE 21
CIRCLE, CHORDS AND ARCS

DEFINITION

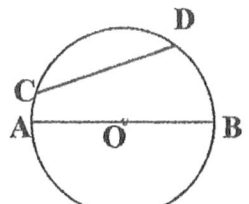

A circle is a figure containing a set of all points equidistant from the fixed point (center).

O is a center of the circle; OB is a radius; AB is a diameter; CD is a chord.

PROPERTIES

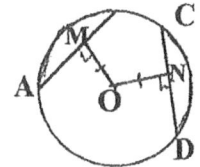

1. Congruent chords intercept congruent arcs.

$$AB = CD \Leftrightarrow \cup AB = \cup CD$$

2. Congruent chords are equidistant from the center.

$$AB = CD \Leftrightarrow OM = ON$$

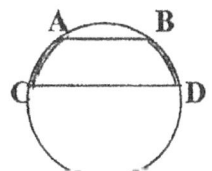

3. Parallel chords intercept congruent arcs.

$$AB \parallel CD \Rightarrow \cup AC = \cup BD$$

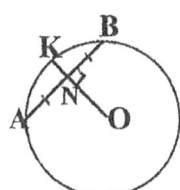

4. A radius perpendicular to a chord bisects this chord.

$$OK \perp AB \Leftrightarrow AN = NB$$

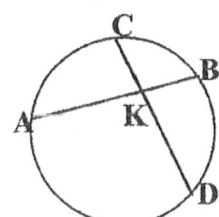

5. The product of the length of two pieces of intersecting chords is constant for all chords with common point.

$$AK \cdot KB = CK \cdot KD$$

TABLE 22
CIRCLE, TANGENTS AND SECANTS

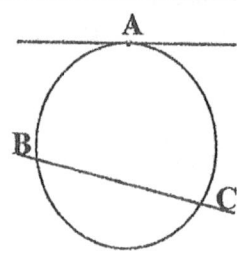

DEFINITIONS

A tangent is a straight line that intersects a circle in exactly one point.

A secant is a straight line that intersects a circle in two points.

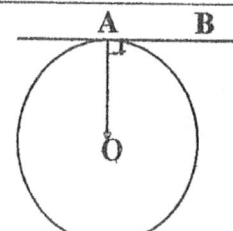

PROPERIES

1. A tangent is perpendicular to the radius at point of tangency.

$$AB \perp OA$$

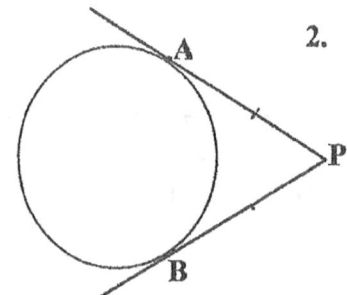

2. Two tangent segments from common point outside of a circle to the point of tangency are congruent.

$$PA = PB$$

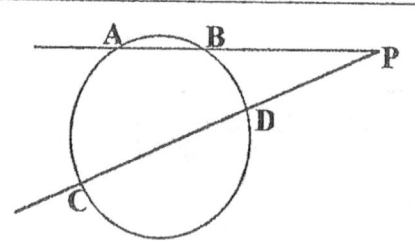

3. The product of the length of a secant segment by the length of its external part is constant for all secants with common point outside of a circle.

$$AP \cdot BP = CP \cdot DP$$

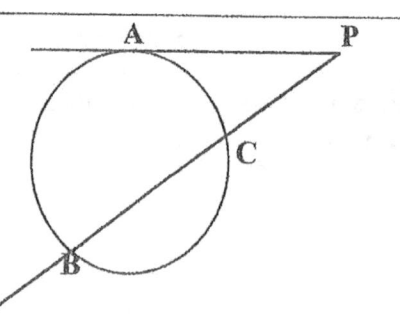

4. If a tangent and a secant have common point outside of a circle, then the length of a tangent segment is a geometric mean of the length of secant segment and its external part.

$$AP^2 = BP \cdot CP$$

TABLE 23
CIRCLES AND ANGLES

CENRTAL ANGLE

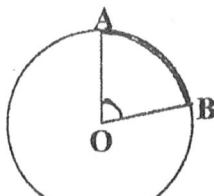

$$\angle AOB = \cup AB$$

$\angle AOB$ is a central angle

INSCRIBED ANGLE

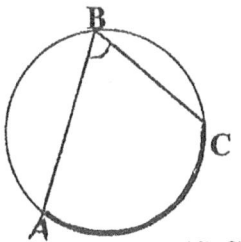

$$\angle ABC = \tfrac{1}{2}\cup AC$$

$\angle ABC$ is an inscribed angle

ANGLE FORMED BY TWO CHORDS

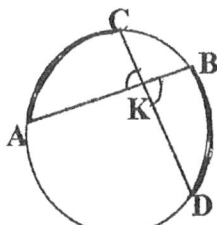

The measure of an angle formed by two chords is equal to half of the sum of two measures of the arcs intercepted by the chords.

$$\angle AKC = \tfrac{1}{2}\,(\cup AC + \cup BD)$$

ANGLE FORMED BY TWO SECANTS

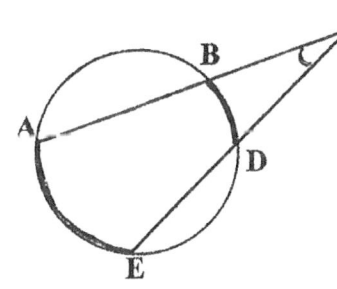

The measure of an angle formed by two secants is equal to half of difference of the measures of two arcs intercepted by the secants.

$$\angle ACE = \tfrac{1}{2}\,(\cup AE - \cup BD)$$

ANGLE FORMED BY TANGENT AND CHORD

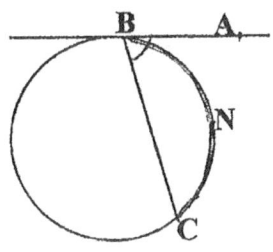

$$\angle ABC = \tfrac{1}{2}\cup BNC$$

ANGLE FORMED BY TWO TANGENTS

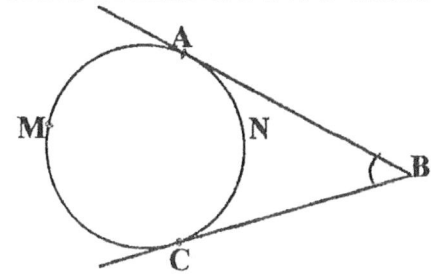

$$\angle ABC = \tfrac{1}{2}\,(\cup AMC - \cup ANC)$$

TABLE 24
INSCRIBED AND DESCRIBED TRIANGLES

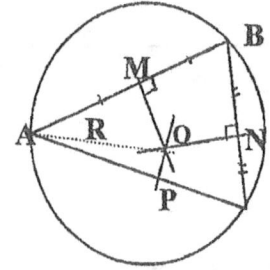

The center of a circumscribed circle is the point of intersection of the perpendicular bisectors of the sides of a triangle.

O is a center;   OA = OB = OC = R

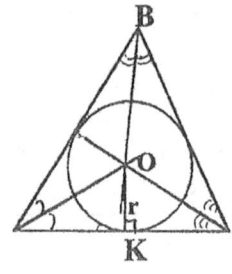

The center of an inscribed circle is the point of intersection of the bisectors of a triangle.

O is a center; OK⊥AC; OK is a radius.

TABLE 25
INSCRIBED AND DESCRIBED QUADRILATERALS

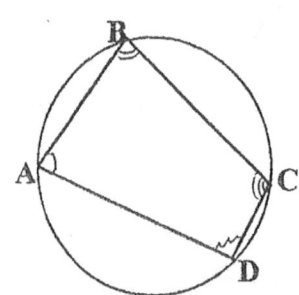

ABCD is inscribed in a circle   ⇔ $\angle A + \angle C = 180°;$
$\angle B + \angle D = 180°$

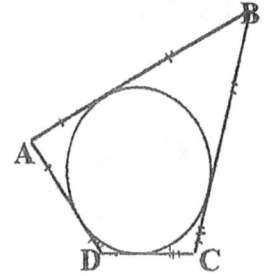

ABCD is described around a circle   ⇔   AB+CD = BC+AD

TABLE 26
PERIMETER AND AREA

TRIANGLE

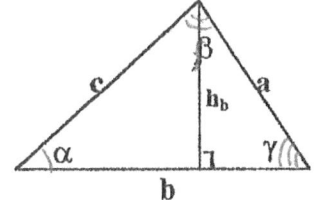

PERIMETER

$$P = a + b + c$$

AREA

1.
$$A = \tfrac{1}{2}\, b h_b$$

2.
$$A = \tfrac{1}{2}\, bc\sin\alpha$$

3. Heron's Formula

$$A = \sqrt{p(p-a)(p-b)(p-c)}$$

where p is a half of the perimeter

4.*
$$A = \tfrac{1}{2} \begin{vmatrix} x_1 & y_1 & 1 \\ x_2 & y_2 & 1 \\ x_3 & y_3 & 1 \end{vmatrix}$$

where (x_1,y_1); (x_2,y_2); (x_3,y_3) are the coordinates of the vertices of a triangle.

5.*
$$A = p\cdot r$$

where
p is half of the perimeter,
r is a radius of the inscribed circle.

6.*
$$A = \frac{abc}{4R}$$

where
a, b, and c are sides,
R is a radius of
the circumscribed circle.

QUADRILATERAL

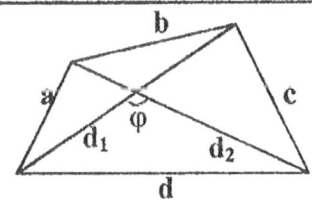

$$A = \tfrac{1}{2} d_1 d_2 \sin\varphi$$

$$P = a + b + c + d$$

PARALLELOGRAM

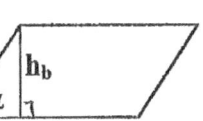

1.
$$A = b h_b$$

2.
$$A = ab\sin\alpha$$

TRAPEZOID

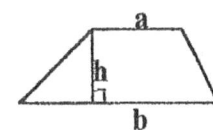

$$A = \tfrac{1}{2}\, h(a + b)$$

CIRCLE AND SECTOR

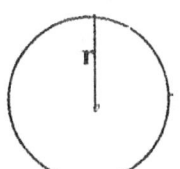

Circumference

Area

$$C = 2\pi r$$

$$A = \pi r^2$$

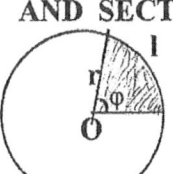

$$l = \varphi r$$

$$A = \tfrac{1}{2}\, \varphi r^2$$

where
φ is in radians.

23

SOLID GEOMETRY

TABLE 27

RELATIVE POSITION OF STRAIGHT LINES

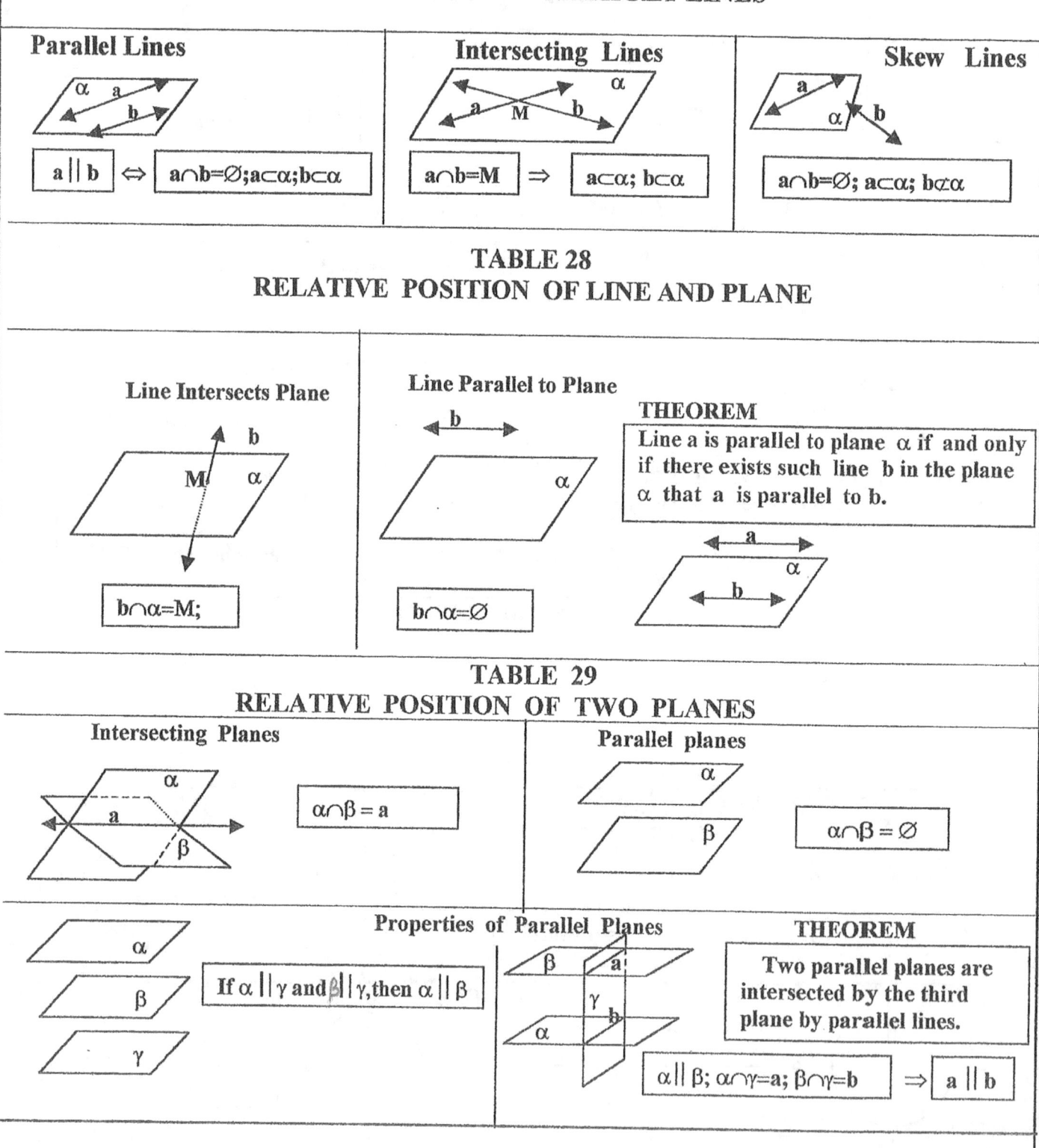

Parallel Lines

$a \parallel b \Leftrightarrow a \cap b = \varnothing; a \subset \alpha; b \subset \alpha$

Intersecting Lines

$a \cap b = M \Rightarrow a \subset \alpha; b \subset \alpha$

Skew Lines

$a \cap b = \varnothing; a \subset \alpha; b \not\subset \alpha$

TABLE 28
RELATIVE POSITION OF LINE AND PLANE

Line Intersects Plane

$b \cap \alpha = M;$

Line Parallel to Plane

$b \cap \alpha = \varnothing$

THEOREM

Line a is parallel to plane α if and only if there exists such line b in the plane α that a is parallel to b.

TABLE 29
RELATIVE POSITION OF TWO PLANES

Intersecting Planes

$\alpha \cap \beta = a$

Parallel planes

$\alpha \cap \beta = \varnothing$

Properties of Parallel Planes

If $\alpha \parallel \gamma$ and $\beta \parallel \gamma$, then $\alpha \parallel \beta$

THEOREM

Two parallel planes are intersected by the third plane by parallel lines.

$\alpha \parallel \beta; \alpha \cap \gamma = a; \beta \cap \gamma = b \Rightarrow a \parallel b$

TABLE 30
LINE PERPENDICULAR TO PLANE

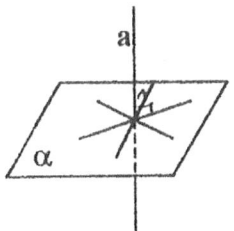

DEFINITION

| $a \perp \alpha$ | $\Leftrightarrow$ | $a \perp x$, where x is any line in α. |

THEOREM

If a line is perpendicular to two intersecting lines in a plane, then the line is perpendicular to this plane.

TABLE 31
ANGLES IN SPACE

ANGLE BETWEEN LINE AND PLANE

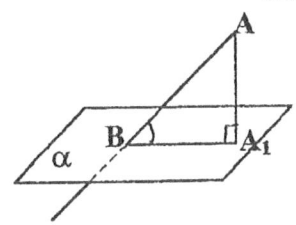

DEFINITION

An angle between a line and a plane is the angle between the line and its projection on the plane.

$\angle ABA_1$ is an angle between line AB and plane α

ANGLE BETWEEN TWO PLANES

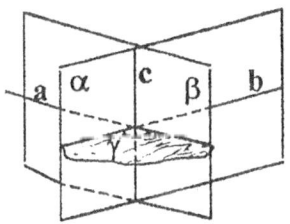

$\alpha \cap \beta = c$; $c \perp \gamma$;
$\alpha \cap \gamma = b$; $\beta \cap \gamma = a$.

$$\angle(\alpha,\beta) = \angle(a,b)$$

$$0° \leq \angle(\alpha,\beta) \leq 90°$$

DIHEDRAL ANGLE

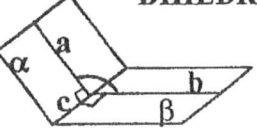

$\alpha \cap \beta = c$; $a \perp c$; $b \perp c$

$$\angle(\alpha,\beta) = \angle(a,b)$$

Dihedral angle is an angle between the two half planes.

PERPENDICULAR PLANES

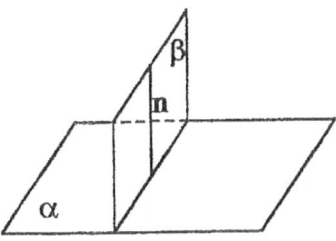

THEOREM

Two planes are perpendicular if and only if one of them Contains a line perpendicular to another plane.

$$\{ n \subset \beta; \ n \perp \alpha \} \Rightarrow \beta \perp \alpha$$

TABLE 32
POLYHEDRA

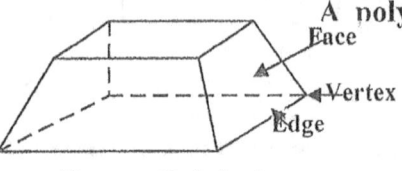

A polyhedron is a simple closed space figure bounded by polygons.

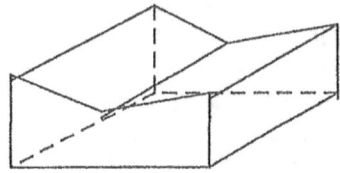

Convex Polyhedron   Concave Polyhedron

EULER'S FORMULA (for convex polygons): $\boxed{F + V - E = 2}$ where
F is the number of faces,
V is the number of vertices,
E is the number of edges.

TABLE 33
PRISMS

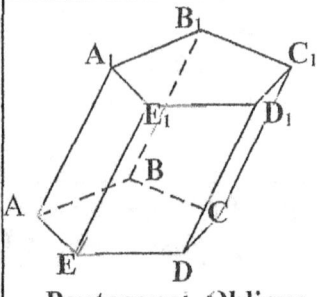

$$\boxed{ABCDE \cong A_1B_1C_1D_1E_1}$$

$$\boxed{AA_1 \parallel BB_1 \parallel CC_1 \parallel DD_1 \parallel EE_1}$$

Pentagonal Oblique Prism

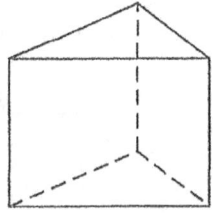

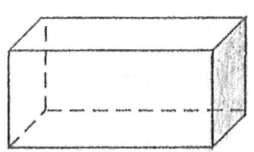

Right Rectangular Prism

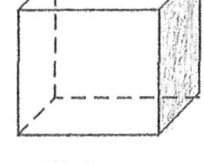

Cube

Right Triangular Prism

SURFACE AREA OF A PRISM : $\boxed{SA = 2A_{base} + LA}$ where
LA is the lateral area.

VOLUME OF A PRISM: $\boxed{V = A_{base}H}$ where
H is the height of the prism.

TABLE 34
PYRAMIDS

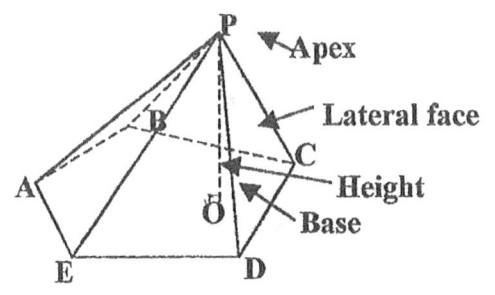

Oblique Pentagonal Pyramid

Right Square Pyramid

H is a height of the pyramid

h is a slant height (apophem)

SURFACE AREA OF A PYRAMID: $SA = A_{base} + LA$ where LA is a lateral area.

VOLUME OF A PYRAMID: $V = \frac{1}{3} A_{base}H$

TABLE 35
REGULAR POLYHEDRA

There exist only next five types of Regular Polyhedra

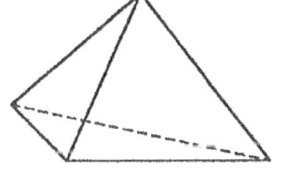   Face
   Face
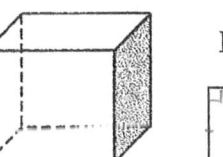
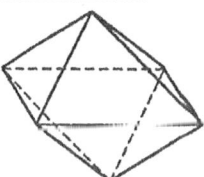   Face

Tetrahedron (4 faces)   Cube (hexahedron, 6 faces)   Octahedron (8 faces)

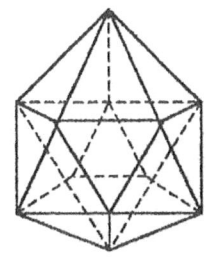   Face

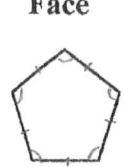   Face

Icosahedron (20 faces)   Dodecahedron   (12 faces)

27

BODIES OF ROTATION
TABLE 36
CYLINDER

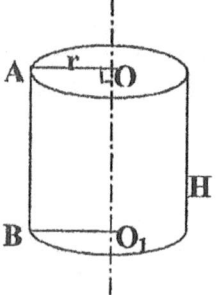

Right Circular Cylinder is a space figure formed by rotation of rectangle around its side.

Cylinder Surface

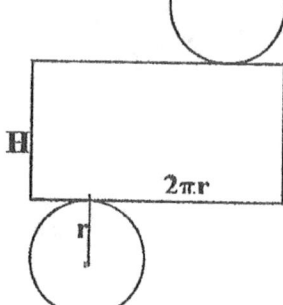

Surface Area

$$SA = 2\pi r^2 + 2\pi rH$$

Volume

$$V = \pi r^2 H$$

TABLE 37
CONE

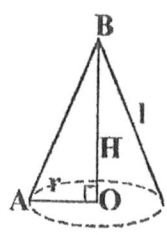

Right Circular Cone is a space figure formed by rotation of a right triangle around its side.

Surface of Cone

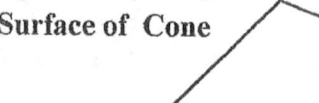

Surface Area

$$SA = \pi r^2 + \pi rl$$

Volume

$$V = \frac{1}{3}\pi r^2 H$$

CONIC SECTIONS

Circle	Ellipse	Parabola	Hyperbola

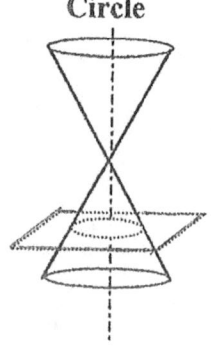
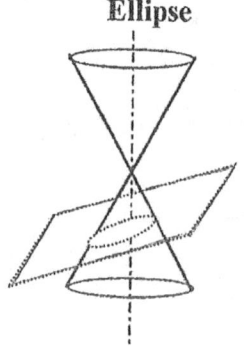

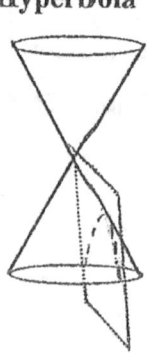

TABLE 38
SPHERE

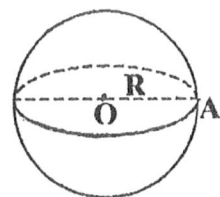

A sphere is a space figure formed by rotation of a semicircle around the diameter.

Surface Area

$$SA = 4\pi R^2$$

Volume

$$V = \frac{4}{3}\pi R^3$$

INTERSECTION SPHERE AND PLANE

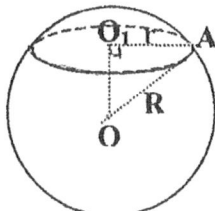

Any section of a sphere by a plane is a circle.

$O_1A = r$ (radius of section); $OA = R$ (radius of sphere); OO_1 is a distance from the center of sphere to the section.

SECTION OF SPHERE BY TWO PLANES

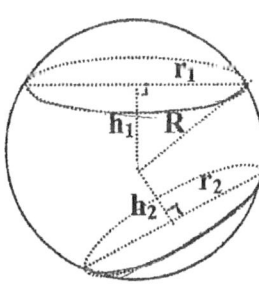

R is a radius of sphere; r_1 and r_2 are radii of sections; h_1 and h_2 are distances from the center of sphere to the planes.

$h_1 = h_2$	$\Leftrightarrow$	$r_1 = r_2$
$h_1 > h_2$	$\Leftrightarrow$	$r_1 < r_2$
$h_1 < h_2$	$\Leftrightarrow$	$r_1 > r_2$

Equidistant from the center of a sphere sections have the same radii. A section which is farther from the center has less radius.

COORDINATE GEOMETRY
TABLE 39
CARTESIAN COORDINATES

On plane	In space
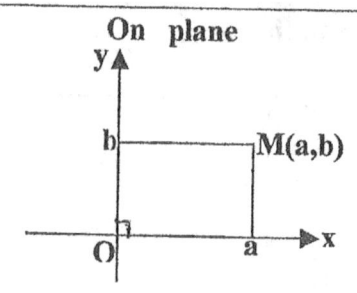	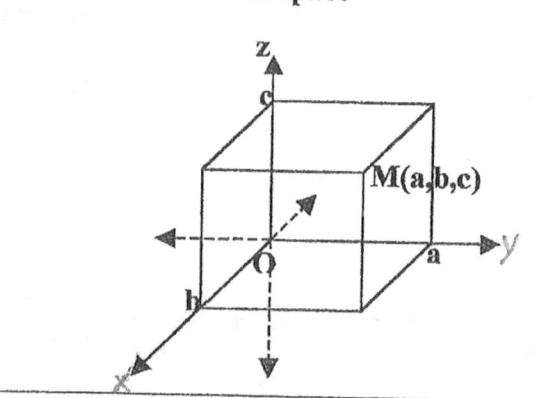

Coordinates of Middle Point of a Segment

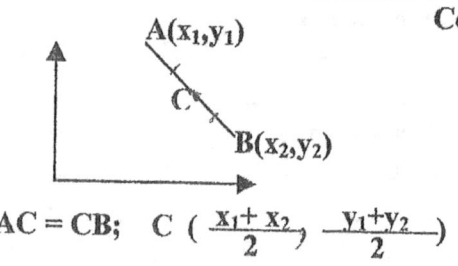

$AC = CB;$ $C \left(\dfrac{x_1 + x_2}{2}, \dfrac{y_1 + y_2}{2} \right)$

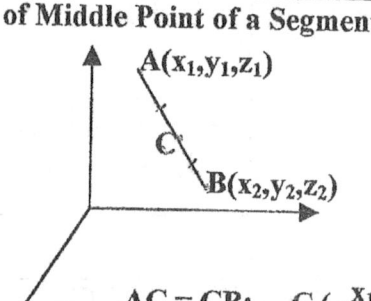

$AC = CB;$ $C \left(\dfrac{x_1 + x_2}{2}, \dfrac{y_1 + y_2}{2}, \dfrac{z_1 + z_2}{2} \right)$

Distance Between Two Points

$AB = \sqrt{(x_2 - x_1)^2 + (y_2 - y_1)^2}$ $AB = \sqrt{(x_2 - x_1)^2 + (y_2 - y_1)^2 + (z_2 - z_1)^2}$

EQUATIONS
TABLE 40
STRAIGHT LINE ON PLANE

Slope of line

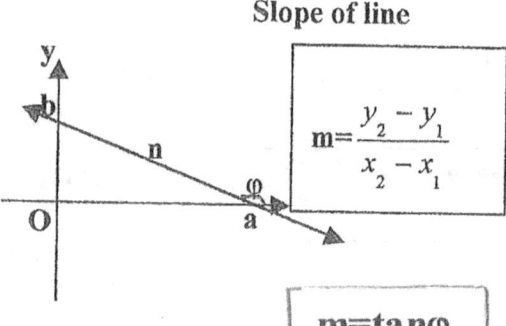

$m = \dfrac{y_2 - y_1}{x_2 - x_1}$

$m = \tan\varphi$

1. Standard form:

$Ax + By + C = 0$

3. Slope-Point form:

$y - y_0 = m(x - x_0)$

2. Slope- Intercept form:

$y = mx + b$

4. Intercepts form:

$\dfrac{x}{a} + \dfrac{y}{b} = 1$

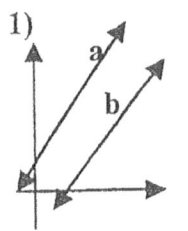

1)

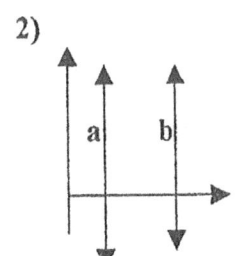

2)

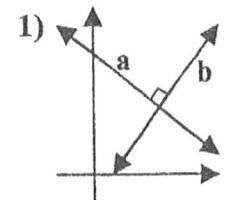

1)

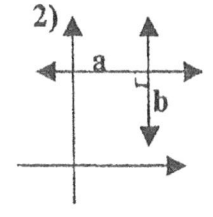

2)

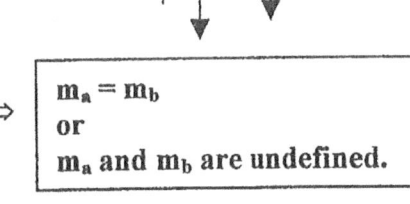

$$\boxed{a \parallel b} \Leftrightarrow \boxed{\begin{array}{l} m_a = m_b \\ \text{or} \\ m_a \text{ and } m_b \text{ are undefined.} \end{array}}$$

$$\boxed{a \perp b} \Leftrightarrow \boxed{\begin{array}{l} m_a \cdot m_b = -1 \\ \text{or} \\ m_a = 0 \text{ and } m_b \text{ is undefined.} \end{array}}$$

TABLE 41
EQUATION OF PLANE IN SPACE

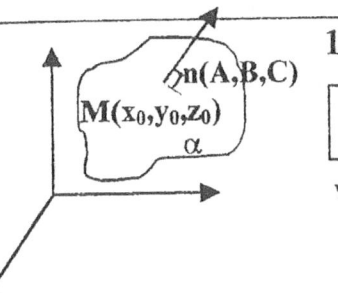

1. Standard form:

$$Ax + By + Cz + D = 0$$

where $n \perp \alpha$, and A,B,C are the coordinates of vector n.

2. Perpendicular Vector- Point form:

$$A(x - x_0) + B(y - y_0) + C(z - z_0) = 0$$

where x_0, y_0, z_0 are coordinates of the given point M.

3. Intercepts form:

$$\frac{x}{a} + \frac{y}{b} + \frac{z}{c} = 1$$

where
a,b,c are points of intersection of the plane with coordinate axises.

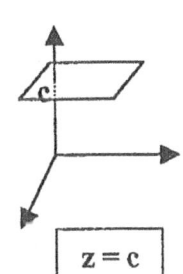

$$\boxed{z = c}$$

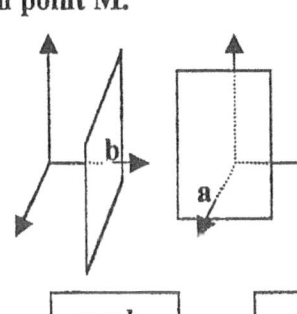

$$\boxed{y = b} \qquad \boxed{x = a}$$

Distance from point to line in plane

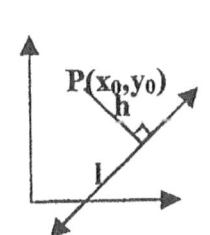

$$h = \left| \frac{Ax_0 + By_0 + C}{\sqrt{A^2 + B^2}} \right|$$

Distance from point to plane

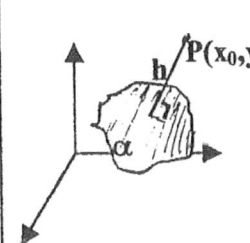

$$h = \left| \frac{Ax_0 + By_0 + Cz_0 + D}{\sqrt{A^2 + B^2 + C^2}} \right|$$

TABLE 42
EQUATIONS OF CIRCLE AND SPHERE

Equation of a circle	Equation of a sphere
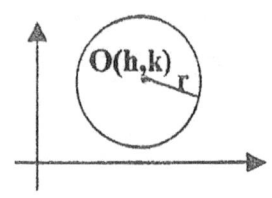	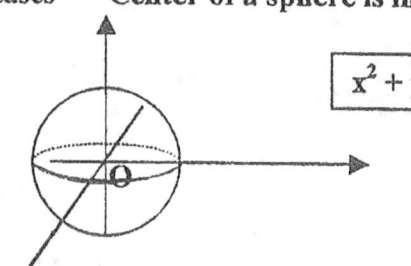
$$(x-h)^2 + (y-k)^2 = r^2$$	$$(x-h)^2 + (y-k)^2 + (z-l)^2 = R^2$$
where	where
h,k are the coordinates of the center of a circle	h,k,l are the coordinates of the center of a sphere.

Particular cases

Center of a circle is in origin	Center of a sphere is in origin
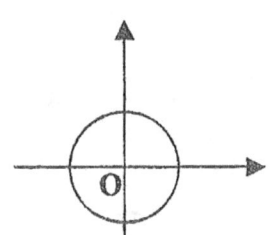	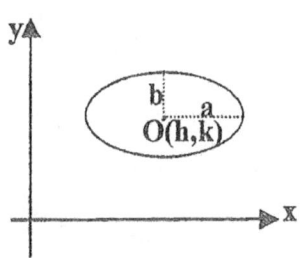
$$x^2 + y^2 = r^2$$	$$x^2 + y^2 + z^2 = R^2$$

TABLE 43
EQUATIONS OF CONIC SECTIONS

Ellipse	Parabola	Hyperbola
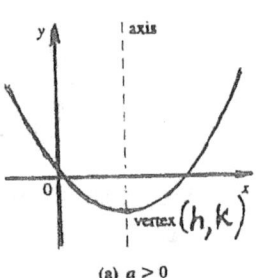		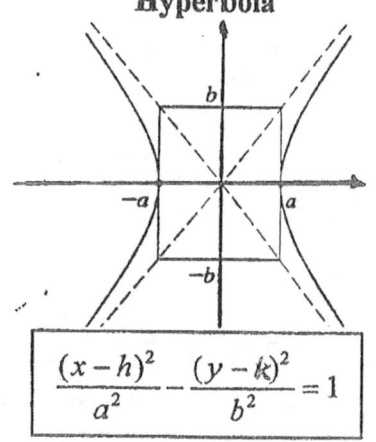
(a) $a > 0$		
$$\frac{(x-h)^2}{a^2} + \frac{(y-k)^2}{b^2} = 1$$	$$y = a(x-h)^2 + k$$	$$\frac{(x-h)^2}{a^2} - \frac{(y-k)^2}{b^2} = 1$$
where	where	where
(h,k) is the center of ellipse, a,b are the halfaxes of ellipse.	(h,k) is the vertex of a parabola, a is a coefficient.	(h,k) is the center of symmetry, a and b are the numbers, such that $\pm\dfrac{b}{a}$ are the slopes of the asymptotes.

VECTORS

TABLE 44
DEFINITION OF VECTORS

DEFINITION

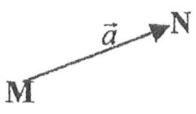

| A vector is a directed segment. |

| The magnitude of vector is the length of the segment. |

$$|\vec{a}| = MN$$

COORDINATES OF VECTORS

On plane

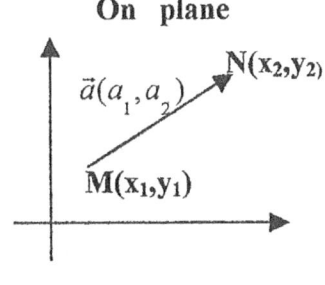

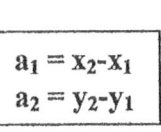

$$a_1 = x_2 - x_1$$
$$a_2 = y_2 - y_1$$

$$|\vec{a}| = \sqrt{a_1^2 + a_2^2}$$

In space

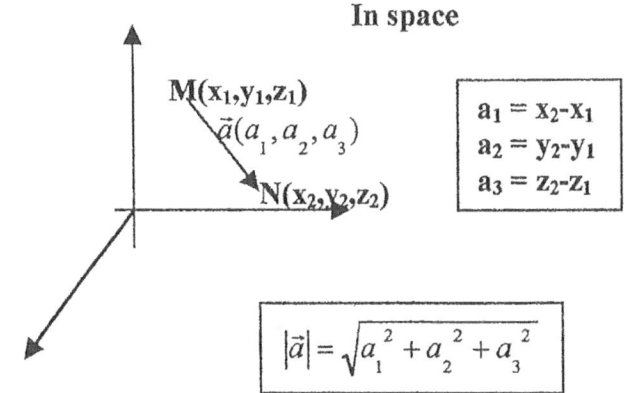

$$a_1 = x_2 - x_1$$
$$a_2 = y_2 - y_1$$
$$a_3 = z_2 - z_1$$

$$|\vec{a}| = \sqrt{a_1^2 + a_2^2 + a_3^2}$$

CONGRUENT VECTORS

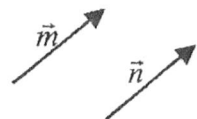

$$\boxed{\vec{m} \cong \vec{n}} \Leftrightarrow \begin{cases} |\vec{m}| = |\vec{n}| \\ \vec{m} \text{ and } \vec{n} \text{ have the same direction} \end{cases}$$

or

$$\vec{m}(m_1, m_2) \cong n(n_1, n_2) \Leftrightarrow \langle m_1 = n_1, m_2 = n_2 \rangle$$

$$\vec{m}(m_1, m_2, m_3) \cong \vec{n}(n_1, n_2, n_3) \Leftrightarrow \langle m_1 = n_1, m_2 = n_2, m_3 = n_3 \rangle$$

33

TABLE 45
VECTOR OPERATIONS

ADDITION OF VECTORS

Rule of a triangle

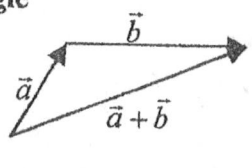

Rule of a parallelogram

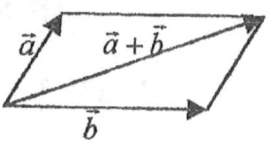

Coordinate form

On plane

$$\vec{a}(a_1, a_2) + b(b_1, b_2) = \vec{c}(a_1 + b_1, a_2 + b_2)$$

In space

$$\vec{a}(a_1, a_2, a_3) + \vec{b}(b_1, b_2, b_3) = \vec{c}(a_1 + b_1, a_2 + b_2, a_3 + b_3)$$

SUBTRACTION OF VECTORS

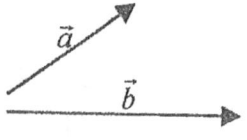

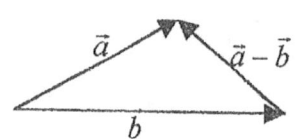

$$\vec{a}(a_1, a_2) - \vec{b}(b_1, b_2) = \vec{c}(a_1 - a_2, b_1 - b_2)$$

$$\vec{a}(a_1, a_2, a) - \vec{b}(b_1, b_2, b_3) = \vec{c}(a_1 - b_1, a_2 - b_2, a_3 - b_3)$$

MULTIPLICATION OF VECTOR BY NUMBER

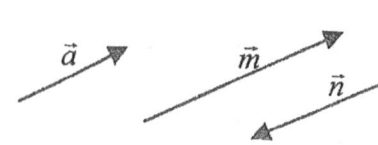

$$\vec{m} = \lambda \cdot \vec{a}, \lambda > 0$$

$$\vec{n} = \lambda \cdot \vec{a}, \lambda < 0$$

$$|\lambda \cdot \vec{a}| = |\lambda| \cdot |\vec{a}|$$

multiplication of a vector by a positive number is a vector directed in the same direction as the original vector.
The result of multiplication of a vector by a negative number is a vector directed in opposite direction to the original vector.

$$\lambda \cdot \vec{a}(a_1, a_2) = \vec{c}(\lambda \cdot a_1, \lambda \cdot a_2)$$

$$\lambda \cdot \vec{a}(a_1, a_2, a_3) = \vec{c}(\lambda \cdot a_1, \lambda \cdot a_2, \lambda \cdot a_3)$$

COLLINEAR VECTORS

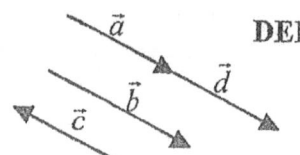

DEFINITION

Vectors are collinear if they belong to the same or parallel lines.

$\vec{a}$ and $\vec{b}$ are collinear $\Leftrightarrow$ $\vec{b} = \lambda \cdot \vec{a}$

SCALAR PRODUCT OF VECTORS

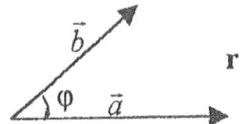
r

$$\vec{a} * \vec{b} = |\vec{a}| \cdot |\vec{b}| \cdot \cos\varphi$$

A scalar product of two vectors is a number.

Coordinate form

$$\vec{a}(a_1, a_2) * \vec{b}(b_1, b_2) = a_1 \cdot b_1 + a_2 \cdot b_2$$

$$\vec{a}(a_1, a_2, a_3) * \vec{b}(b_1, b_2, b_3) = a_1 \cdot b_1 + a_2 \cdot b_2 + a_3 \cdot b_3$$

Corollary

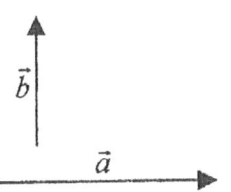

THEOREM

For $|\vec{a}| \neq 0$ and $|\vec{b}| \neq 0$

$$\vec{a} * \vec{b} = 0 \Leftrightarrow \vec{a} \perp \vec{b}$$

TABLE 46
DECOMPOSITION OF VECTOR

On plane

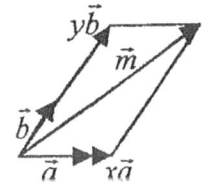

THEOREM

For any $\vec{m}$ and noncollinear $\vec{a}$ and $\vec{b}$ there exist such unique real numbers x and y that $\vec{m} = x\vec{a} + y\vec{b}$

In space

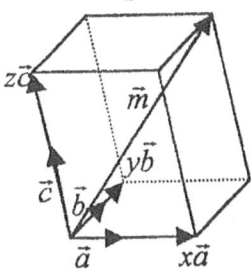

THEOREM

For any $\vec{m}$ and noncoplanar $\vec{a}, b$ and $\vec{c}$ there exists such unique real numbers x,y and z that $\vec{m} = x\vec{a} + y\vec{b} + z\vec{c}$

Particular Case : Decomposition by Unit Coordinate Vectors

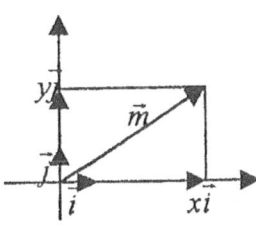

$$\vec{m} = x\vec{i} + y\vec{j}$$

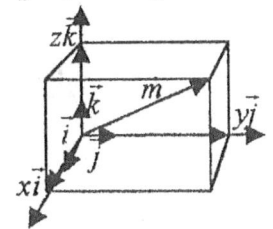

$$\vec{m} = x\vec{i} + y\vec{j} + z\vec{k}$$

TRANSFORMATIONS
TABLE 47
ISOMETRY

Transformation is such displacement of a figure when there is one-to one correspondence between each point in original figure and its image.

An isometry is a transformation that preserves distance

TRANSLATION

Properties of translations

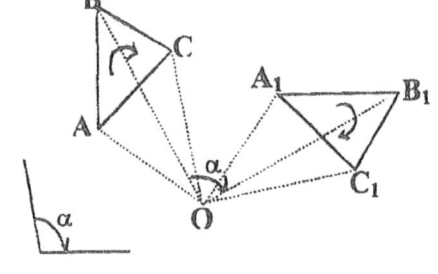

$$\vec{AA_1} \cong \vec{BB_1} \cong \vec{CC_1} \cong \vec{n}$$

1. Translations take lines to lines, rays to rays, line segment to line segments.

2. Translations preserve distance, angle measure, perpendicularity, parallelism, and orientation.

ROTATION

Properties of rotations

O is the center of rotation, α is a directed angle.

1. Rotations take lines to lines, rays to rays, line segments to line segments.

2. Rotations preserve distance, angle measure, perpendicularity, parallelism, and orientation.

REFLECTION

Pronerties of reflections

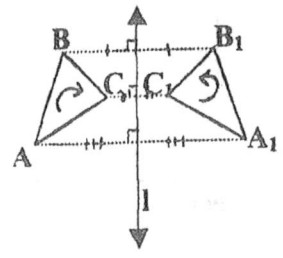

Reflection with respect to line l

1. Reflections take lines to lines, rays to rays, line segments to nine segments.

2. Reflections preserve distance, angle measure, perpendicularity , and parallelism.

3. Reflections change orientation.

Two figures are congruent if and only if there is an isometry that takes one figure to the another.

TABLE 48
SIZE TRANSFORMATION

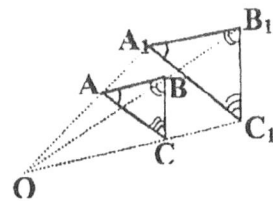

O is the center of size transformation

$$\frac{OA_1}{OA} = \frac{OB_1}{OB} = \frac{OC_1}{OC} = k$$

where k is a coefficient of the size transformation.

$$\Delta A_1 B_1 C_1 \sim \Delta ABC$$

Properties of size transformation

1. Size transformations take lines to lines, rays to rays, line segments to line segments.

2. Size transformations preserve ratio of distances, angle measure, perpendicularity, parallelism, and orientation.

3. Size transformations take lines to parallel lines.

SIMILITUDES

A similitude is a combination if a size transformation and an isometry.

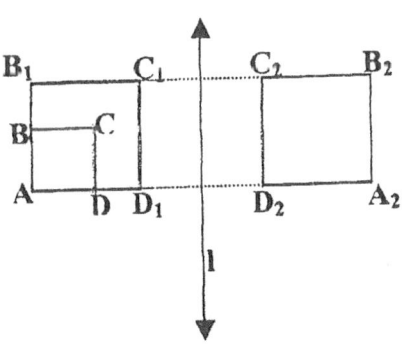

1. $AB_1C_1D_1$ is the image of ABCD under size transformation with center A and k = 2.

2. $A_2B_2C_2D_2$ is the image of $AB_1C_1D_1$ under reflection with respect to line l.

Two figures are similar if and only if there is a similitude that takes one figure to the another.

PROJECTIVE GEOMETRY

TABLE 49
PROJECTION

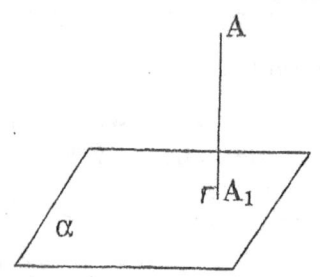

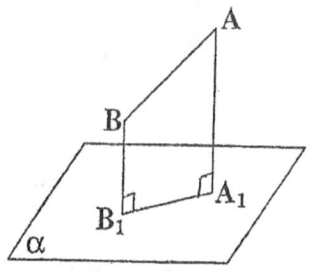

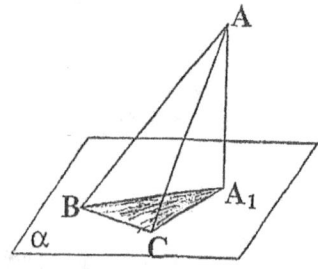

A_1 is the projection of point A onto plane α.

A_1B_1 is the projection of segment AB onto plane α.

$\triangle A_1BC$ is the projection of $\triangle ABC$ onto plane α.

A dual theorem is a theorem that can be obtained from another by interchanging the dual elements - lines and points.

THE PRINCIPLE OF DUALITY OF PROJECTIVE GEOMETRY

If one of two dual theorems is true, then the other theorem is also true.

TABLE 50
DESARGUES'S THEOREM

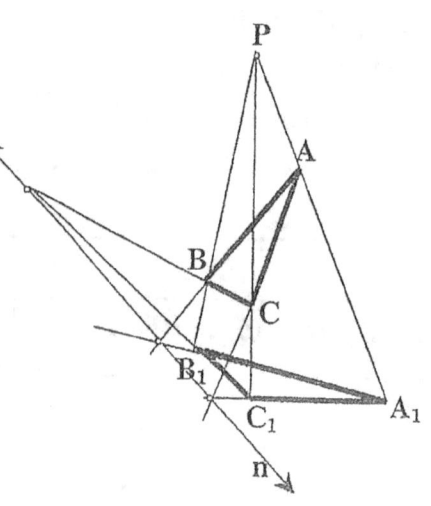

If the lines joining corresponding vertices of two triangles are concurrent (that is, they meet in one point), then the points of intersection of corresponding sides are collinear.

Let $AB \cap A_1B_1 = S$, $BC \cap B_1C_1 = T$, $AC \cap A_1C_1 = Q$

$BB_1 \cap AA_1 \cap CC_1 = P$ $\Leftrightarrow$ $S, T, Q \in n$

TABLE 51
PAPPUS'S THEOREM

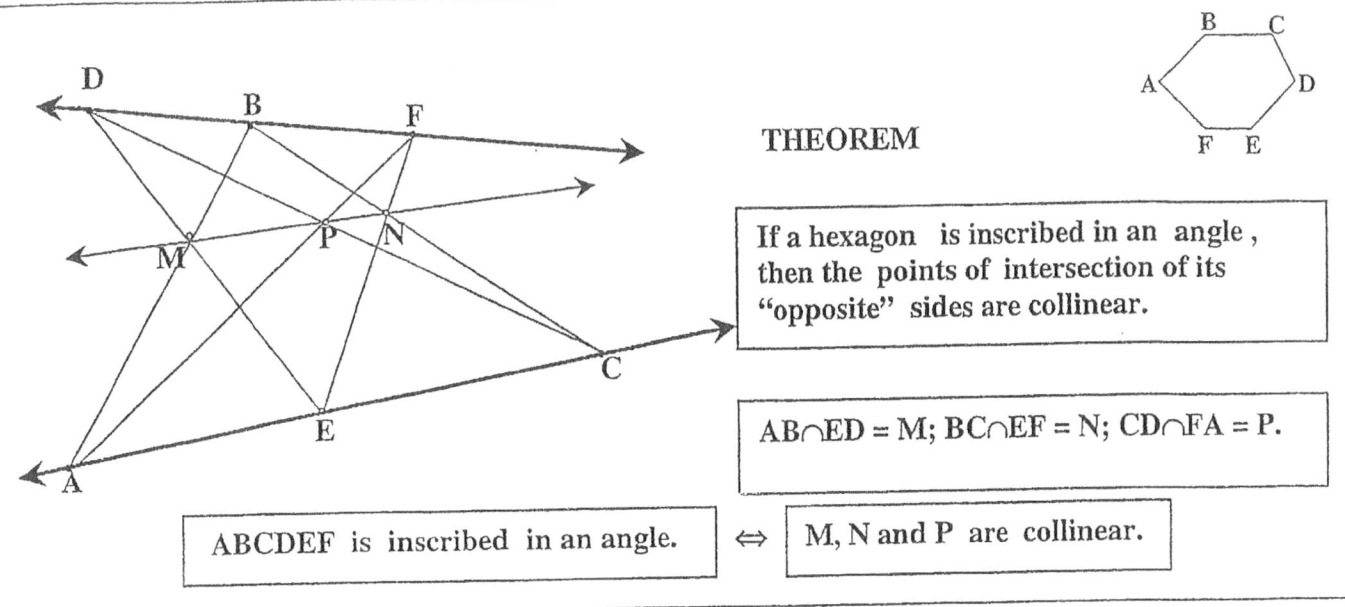

THEOREM

If a hexagon is inscribed in an angle, then the points of intersection of its "opposite" sides are collinear.

AB∩ED = M; BC∩EF = N; CD∩FA = P.

ABCDEF is inscribed in an angle.	⇔	M, N and P are collinear.

TABLE 52
PASCAL'S THEOREM

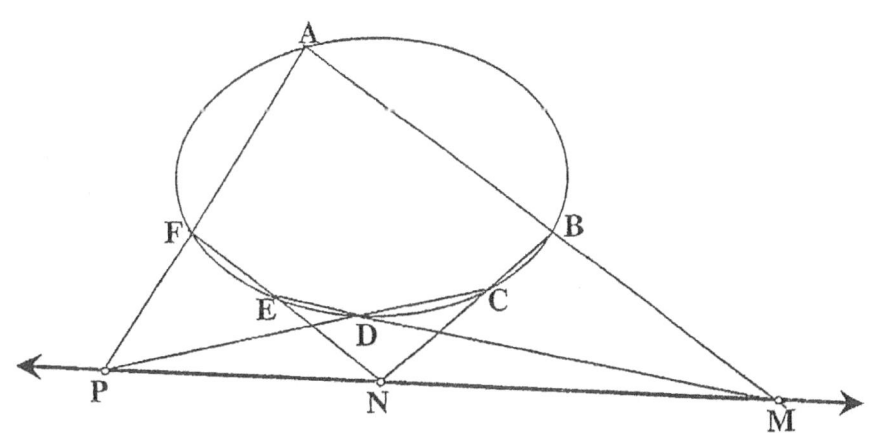

THEOREM

If a hexagon is inscribed in a conic section, the three points of intersection of pairs of opposite sides are collinear.

NON-EUCLIDEAN GEOMETRIES
TABLE 53
PARALLEL POSTULATE

PARABOLIC GEOMETRY

EUCLIDEAN GEOMETRY   (~300 BC)

For every line l and every point P not on line l, there exists a unique line m that contains P and is parallel to l.

HYPERBOLIC GEOMETRY

LOBACHEVSKIAN GEOMETRY   (~ 1830)

For every line l and every point P not on line l, there exist at least two different lines that contain P and are parallel to l.

ELLIPTIC GEOMETRY

RIEMANNIAN GEOMETRY   (~1850)

For every line l and point P not on line l, there is no line containing P and parallel to l.

In 1872 Felix Klein in his famous *Erlanger Programm* made classification of Geometries.

In 1899 David Hilbert, one of the greatest mathematician of 20th century, created complete, independent and consistent system of axioms.

TABLE 54
COMPARING SOME FACTS IN DIFFERENT GEOMETRIES

EUCLIDEAN	LOBACHEVSKIAN	RIEMANNIAN
Geometry on Plane	Geometry on Pseudosphere	Geometry on Sphere
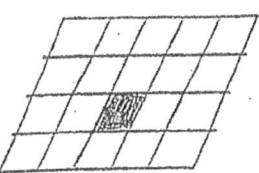	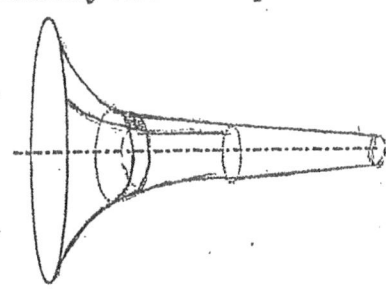	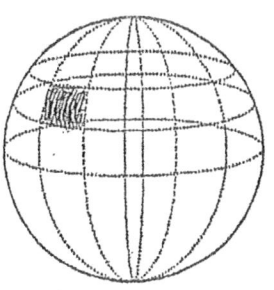

Two points determine a unique line and line segment.

Two points determine at least two line segments.

Triangle

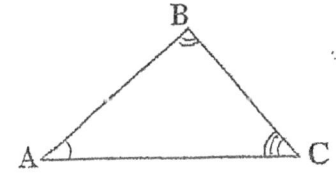

$$\angle A + \angle B + \angle C = 180°$$

Triangle

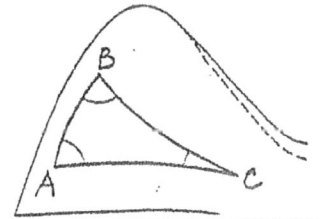

$$\angle A + \angle B + \angle C < 180°$$

Triangle

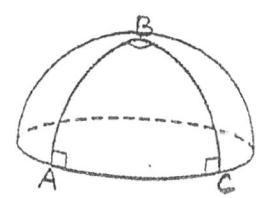

$$\angle A + \angle B + \angle C > 180°$$

Triangles with the same angles might be either congruent or similar.

Triangles with the same angles are congruent.

Straight lines have infinite length.

Straight lines have finite length.

www.ingramcontent.com/pod-product-compliance
Lightning Source LLC
Chambersburg PA
CBHW081237170526
45165CB00009B/3087